FANTASY BASKETBALL

A WINNING PLAYBOOK

Keith Meinelt

authorHOUSE®

AuthorHouse™
1663 Liberty Drive
Bloomington, IN 47403
www.authorhouse.com
Phone: 1-800-839-8640

First published by AuthorHouse 9/8/2009

ISBN: 978-1-4490-1139-0 (e)
ISBN: 978-1-4490-1138-3 (sc)

Printed in the United States of America
Bloomington, Indiana

This book is printed on acid-free paper.

Acknowledgements

I offer my special thanks to Brian Sechrest, who introduced me to fantasy basketball last year and convinced me it might be fun. He was right! Brian is a great sports fan and a fantasy sports veteran. He offered me a great deal of advice with my leagues and with this manual.

Thanks to my wife, Melissa, who helped with the production of this manual. Melissa was my "cheerleader" throughout the project.

Thanks to Bill Lunsford, who created the cover art for this book. Bill's creativity and talent help give the words life, and his visual interpretations inspired me to produce my best work. Bill can be contacted at reddogdesignstudio@yahoo.com.

Thanks to Dr. Arun Inamdar for his humorous illustrations. Arun was a pleasure to work with and I highly recommend his services. You can contact Arun through www.Guru.com, or you can email him at arun.b.inamdar@gmail.com.

Thanks to Lance Shull. Lance introduced me to some of my favorite pastimes, including contract bridge IMPS, Texas Hold'em poker, and stock market investing. Lance is a great competitor, and he has participated in fantasy sports for many years. I'm following Lance's lead yet again.

Thanks to everyone who proofread this work: Ashley Champion, Marilyn Praw, Brian Sechrest, Melissa Buchanan, Greg Boyd, Mike Praw, Jim Raper, Joyce Lee and Katie Gaynor.

Thanks to all my friends and family members who continually asked about the progress of this book and offered suggestions for improving it. These discussions fueled my motivation throughout the process.

Thanks to Yahoo! Yahoo! Fantasy Sports is a top notch environment to "play" in. The graphics are visually pleasing, the mechanics are user friendly, and the Yahoo! community is first rate. I look forward to participating in Yahoo! leagues and "sitting courtside" with my ball players for many seasons to come.

Thanks to Rotoworld. Rotoworld's Draft Guide and Season Pass are outstanding products and great values. Rotoworld's sportswriters provide first rate commentary throughout the season. Their timely posts and emails are an invaluable fantasy sports resource.

I express my great appreciation for the efforts of Team Ziguana at Ziguana.com: Matthew Honea, Bret Morris, Cassidy Morris and Kenny White. Without the ideas of these impressive individuals, and without their hard work and cutting edge website, my rookie fantasy sports season would have been much less interesting and certainly less successful. Ziguana was the driving force behind this project, and I wish Team Ziguana every possible reward. They earned it.

Dedication

To our loving memory of Ernest Gerald Buchanan, Sr.

Gerald fought cancer throughout the 2008-09 NBA season.

We lost Gerald a week before the playoffs began.

We love and miss you, Gerald.

Table of Contents

Introduction

It's 6:00 AM Monday morning. I woke up before my alarm sounded, and in my sleepy state I consider my options. Should I try to catch a little more shut-eye before I prepare for a week at my nine-to-five? Should I eat some breakfast and watch a recorded episode of *Pardon the Interruption?* Decisions, decisions…

6:02 AM: I'm at the computer. No more rest for me, and breakfast can wait because I have more than an hour before I leave for work. That leaves me plenty of time to see how the players from my three fantasy basketball leagues performed last night! This shouldn't take long…

6:45 AM: Wow, there was a lot of NBA activity last night! My players performed very well, and I just located a great free agent. Uh, oh…not much time to clean up and eat breakfast before I leave! So much for spending quality time with my wife and kitty this morning!

Hi everyone. My name is Keith and I am a fantasy sports addict. I'm relatively new to fantasy sports. As I write this manual, my first season of fantasy NBA hoops has come to an end, and my rookie fantasy baseball season is now underway.

I like to believe I'm not a typical fantasy sports rookie. I adapted to the "sport" quickly and my rookie year has been a great success. I won all three of my basketball leagues this past 2008-09 season, and I'm currently in first and second place in my two baseball leagues.

For years I picked on my friends for participating in fantasy sports. I told them repeatedly, "I prefer to live in the *real* world, thanks." Well, no longer. Basketball playoffs have ended and I'm already looking forward to next season's draft. Fortunately I have my baseball leagues to monitor in the meantime. I told you I was an addict.

Before I explain why I wrote this book, let me first tell you what I did not intend to accomplish with this work. I will not describe most of the mechanics involved with online fantasy basketball. It's likely you already participate in fantasy sports and I don't want to waste your time. I'll assume you understand general fantasy sports terminology, and that you have at least a basic understanding of basketball statistics. If you're new to the scene, you'll learn the ins and outs of your fantasy leagues by navigating through your league's website.

I will also not offer advice about which specific players you should draft, which positions tend to be shallow or deep, or how you should set your lineups each day. The NBA is very dynamic and any advice I might offer today can easily be wrong tomorrow. Besides, if you already have experience managing fantasy teams, you likely have just as much (if not more) knowledge in these areas than I have, and it's not my intention to insult you.

What I will attempt in this work is to describe some powerful fantasy sports resources that are available online, and I'll explain how to use these resources to win your leagues. I'll describe the procedures I created that helped me make informed and objective fantasy decisions which led to my success. I will also share the valuable spreadsheet I developed and constantly improved throughout the basketball season. The information I offer in these pages should help you accurately evaluate any given NBA player and help you successfully compete in your leagues.

While drafting this manuscript, I deliberated about who my intended audience should be. Some sections of this manual are targeted at fantasy sports beginners. Other sections should assist the majority of fantasy team owners. Ultimately, everyone can take something from this work, including fantasy sports veterans and even managers of traditional pen-and-paper leagues.

In the end, my objective is simple: I'm offering suggestions for player performance research and analysis. I'm also providing strategy techniques that fantasy managers everywhere can utilize and improve upon. I want to help you use this information to make wise, calculated decisions and to enhance your fantasy sports experiences. In essence, I'm presenting you with my winning playbook.

For Fantasy Baseball Managers

The methods I describe in this book can easily be adapted to fantasy baseball. We're halfway through the 2009 baseball season and I've already tailored my spreadsheet and my player research methods to fit the nuances of fantasy baseball. The transition was virtually "seamless."

There are many differences between the two fantasy sports – some obvious and some subtle – but there are also many similarities. In the end, you can succeed in your basketball leagues *and* in your baseball leagues from reading this manual and incorporating my suggestions.

Though I discuss different fantasy sports websites and resources throughout this book, I am not, and I have never been, affiliated with, or compensated by, any of these companies in any way. I found most of these resources (and many others) on my own during my rookie season, and I'm sharing with you those that I found most helpful. In addition, please know that Rotoworld.com and Ziguana.com are not affiliated with each other in any way.

For your convenience, I repeat all of my procedures, techniques, and spreadsheet information in condensed form in 'The Playbook' and Appendix sections of this manual. In addition, I offer the spreadsheet as a free download at the following website: www.awinningplaybook.com. I hope you enjoy reading this guide as much as I enjoyed writing it!

PRESEASON

Check your team every day…even if you're on vacation… or floating on a small circular piece of ice in the Nordic Sea, assuming you have a cell phone with you. It is your obligation to log in at least once a day, even if only for a minute. Set your lineup properly. Don't leave lackluster, ineffective players on your starting roster just because you're too lazy to hit the waiver wire and find someone better.

Matt Stroup, Rotoworld

I enjoy fantasy sports for many reasons. First and foremost, I'm a sports fan. I enjoy watching sporting events, and I like knowing what cool things are happening in the sporting world. I like to know when athletes break Olympic world records, and I enjoy hearing when an NBA player scores 50 points in a game. If it's sports related, and especially if it's front page news, I can appreciate it.

Another reason I enjoy fantasy sports so much is because I have a competitive nature. I like to compete, and fantasy sports offers a fun and healthy environment for me to do so. I often wager a soda when I compete, and it gives me great pleasure to savor every last drop of my winning drink with the individual who had to buy it for me. Along the same lines, it fuels my fire when I receive friendly, but antagonistic, messages from my competing managers after their players have outperformed mine, especially when their talented players were previously on my roster!

Though we're often drawn to the more popular athletic studs, the stature of the player I acquire and the team he plays for are virtually irrelevant. What's important to me are a player's historical statistics, his current playing status, and what I believe his future performance level will be. In the end, little else matters. It's fun to perform this research, and it's very rewarding when a free agent I've located breaks out as a star. Regardless of the amount of time you can invest in your leagues, what I offer in these pages are methods you can utilize to help you make the most of your research time.

Please know that if you follow my advice, you will need to be an active team manager. I don't mean to imply that you should spend as much time as I did analyzing information. For your sanity's sake, it's probably best you don't! I just mean that by the time the fantasy NBA season ends, if you follow my suggestions, you will be familiar with the performance levels of most NBA players, and you might execute more add/drop transactions than your competing managers.

Assuming you will be an active owner, you should join leagues that do not limit your transactions (i.e. player adds and drops). If your league commissioner significantly limits your available transactions, you could suffer from a lack of flexibility. This is especially true if you join a league that includes end-of-season playoffs, like many H2H leagues include. I'll explain why in 'The Playoffs' section of this book.

Trades are a different matter. Don't worry if your league commissioner limits your trades. In small leagues (100 draft picks or fewer) or in standard-sized leagues (150 draft picks), one or two well-timed and carefully calculated trades should be enough to help you win. Later I'll discuss why I believe it's best *not* to spend lots of time calculating "perfect" trade offers. Most of our time can be better spent effectively analyzing and acquiring the right free agents at the right times instead of trying to manipulate other managers' rosters.

I also suggest not joining very deep leagues (200+ draft picks). In deep leagues, most of the fantasy-valuable NBA players will be owned and the free agent selection will be limited. In such leagues, trades are more important. That's fine if the other team owners remain active and regularly monitor your league's activities. However, as any fantasy veteran will tell you, some managers will invariably neglect their fantasy managerial duties, and trade offers to these inactive managers will be ignored. As a result, your time-consuming trade research will be wasted. Stay with small or standard-sized leagues and focus on acquiring quality free agents to avoid these trade-related headaches.

Regarding your league participation, my suggestion is to join one rotisserie (Roto) league and one head-to-head (H2H) league. You'll have fun experiencing the different nuances that each scoring style offers. If you join too many leagues, you can spread yourself too thin. On the other hand, if you remain active throughout the season, but you join only one league, you might burn out from analyzing the same data over and over again. Even worse, you might woodenly execute transactions from boredom, which is a waste of your precious time and can also be counterproductive.

This year I've spent many hours researching fantasy sports, evaluating player performances and monitoring my leagues. Fortunately for you, I've determined what works and what doesn't. Regarding your time required, I'll offer a subjective guideline: allow at least three hours per week, per league. If you join a league with season-ending playoffs, allow at least thirty minutes per day during the playoffs. I spent much more time than this managing my first three leagues during my rookie season (just ask my family and friends), but I became more efficient as the season progressed, and you can benefit from the streamlined procedures I offer in this pages.

THE DRAFT

Drafting is overrated, but it is still very important in determining your team's final outcome. Winning managers put as much effort into their draft as they do the free agency and waiver wire throughout the season. The trick to coming out on top is to be knowledgeable before the season starts and to remain active throughout the season.

Bret Morris, Ziguana.com

The draft, as important as it is, is probably overrated. Consider my case as supporting evidence. In my three leagues combined, I drafted a total of 39 players (13 players per team). By the season's end, I retained only 12 of these original 39 players, and seven of them were listed on Yahoo!'s "Can't Cut" list. The reason I dropped so many of my draft picks is simply because their performances dropped to relatively low levels for extended periods of time.

As an (overly) active manager, I admittedly dropped many of my drafted players too soon. I will argue, however, that it was smart to eventually drop most, if not all, of the players I dropped, which leads me to my point: don't sweat the draft. A successful draft is only one aspect that can help lead you to fantasy victory. With that said, the better you draft, the fewer transactions will be required throughout the season. Therefore, it certainly helps to pay attention during the pre-season and to research the experts' views and observations as draft day approaches.

Earlier, I mentioned that I wouldn't tell you which players to draft. While I won't give you specific names, what I will tell you is to follow the experts' advice on draft day. There are many sources of information available, both in print and online, that can reliably assist you with your draft selections. Personally, I think Rotoworld's Draft Guide is a great source for this information. If you use Rotoworld's Draft Guide, enter your league settings on the website's 'Custom Scoring' section. Then, exactly one day before the draft, print out your customized cheat sheet.

On your league's draft webpage, take the time to electronically pre-rank every single player who is expected to have fantasy value in your league. If you join a league with 8 managers and each team will draft ten players, then pre-rank at least 80 players at your fantasy league's website. If you join a league with 12 managers and every team will draft 13 players, pre-rank at least 156 players. If you join multiple leagues within the same host site (in my case, Yahoo!), you should be able to copy your pre-rankings from one league to the next.

As tempting as it might be to rely on last year's data alone to create your draft cheat sheet, I wouldn't recommend it. Things change constantly throughout the year, including the offseason. The expert fantasy sportswriters know their stuff, and it's best to follow their advice regarding the draft. Steve Alexander of Rotoworld.com claims, "Rotoworld's Draft Guide helps make even the novice appear knowledgeable," and as a former novice, I couldn't agree more.

Assuming you have pre-ranked your players at your league's website, you *could* skip the actual draft and allow the league's computer system to auto-draft your players for you. For shame! As much fun as I had during the regular season and the playoffs, the action-packed draft is a can't-miss event, especially if you join leagues with personal friends. Don't deprive yourself of this entertainment. Don't auto-draft. Live draft only! In the end, if you do nothing more than pre-rank every player based on expert opinions, and you choose the top available player from your custom list during each round of the draft, you'll do fine.

Immediately following the draft, review your team and look for weaknesses. Don't wait until the next week or even the next day; do it right away. If you mistakenly drafted a player who you believe you should not have drafted, search the waiver wire and enter your request for an appropriate replacement player. The more competitive your league is, the more likely other managers will be doing the same thing, and it's important you enter your request for your target player before someone else does.

During the days after the draft, but before the season begins, check your league periodically and review any activity. Research your competing owners and try to identify the stronger managers in your league. At your league's website, you should be able to view each manager's historical fantasy sports results. If you identify managers who you believe will be competitive, monitor their actions throughout the season. Pay attention to what these managers do and try to determine *why* they make the decisions they make. Every advantage you can gain will help as you approach opening day.

OPENING DAY

Fantasy managers rely on our site for timely, accurate news. Owners adjust their rosters based on what writers at Rotoworld tell them. Many people don't know how much digging we do to obtain the news. By contacting the right resources, we can find out before game time if a player who is listed as 'questionable' is going to play. We work hard to find out what's going to happen before it happens.

Steve Alexander, Rotoworld

Opening day is great. The drama of the draft is over and we begin to envision a season of fantasy success. If our last-round pick from the draft scores twenty points and grabs ten rebounds on his first night out, we can rejoice in our own good fortune. If we find ourselves at the top of our league after only a day or two, we can wonder in amazement at our own drafting genius, and we can start posting smack talk on our leagues' message boards. Don't be cocky, though, because the season is long and full of surprises!

As far as my opening day advice is concerned, I'm reminded of a pivotal scene from the classic sci-fi movie, *Total Recall*. Douglas Quaid's character views his real personality, Hauser, on the computer screen and he hears himself say, "Now, this is the plan: get you're a__ to Mars." With this scene in mind, I say to you… Now, this is the plan: get yourself to Rotoworld's Season Pass website.

The Season Pass site includes many excellent features that help guide fantasy managers throughout the season. In the 'Configuration Settings' section, enter any player that appears on one your rosters. Set up your email account in the 'Roto Direct' section and enter your request for Rotoworld to email you updates on your players. These updates will continuously keep you informed of what you need to know to manage and maintain healthy and productive rosters. You might also request the company's other feature articles like *The Daily Dose*, *Roundball Stew* and *Waiver Wired*. By simply reviewing your daily emails from Rotoworld, you can be confident you'll remain up to date with all the pertinent happenings of the NBA.

1ST QUARTER

The z-score essentially normalizes each statistic by representing how many standard deviations a player is away from the average in that category. At Ziguana, we then tweak our formulas to take into account things like statistical volume for certain categories, which is extremely important for fantasy managers. Ultimately, these z-scores can be used to calculate and compare the overall values of players across any number of categories in roto or head-to-head leagues.

Kenny White, Ziguana.com

In November 2008 I searched the Internet for additional NBA-related resources. I was looking for sources of information that provided relevant and customizable data that would help me evaluate player performances. I found many websites that listed extensive game log records, but this data wasn't sufficient because too much time was required to accurately compare players' performance levels in specific categories and for specific time periods.

Some websites offered impressive information, but either the sites were not user friendly enough, or they didn't allow users to customize the data as much as I wanted. Eventually I discovered a website that gained my attention immediately, and the more I navigated the site, the more I became hooked. Enter the world of...

www.Ziguana.com

Ziguana's main focus is on player and league analysis for the NBA and MLB, and the website's designers present the data in a unique and truly customizable format. Ziguana allows us to customize our research to suit our specific league settings and scoring formats. In addition, Ziguana allows us to effortlessly oversee any Yahoo! or ESPN fantasy leagues. It's not my intention to describe all the features of this website, but I will offer a few reasons why I'm promoting this resource.

First, as stated in the quote at the beginning of this section, Ziguana incorporates *volume* into its ratio category formulas. Ziguana assigns a "z-score" to each statistical category for every player, and volume is a critical component of ratio categories. There are many ratio categories in the NBA. The most common ones include FG% (field goal percentage), FT% (free throw percentage), 3PT% (three point percentage) and A/T (assist to turnover ratio). Consider the following example using FT%:

| Player A | → | FT%: 95% | FT% z-score: | 0.50 |
| Player B | → | FT%: 85% | FT% z-score: | 1.00 |

If FT% is our only consideration, which player do we want on our team? At first glance, it appears Player A is the better choice because he successfully converts a higher percentage of free throws (95%) than Player B converts (85%). In other words, Player A is the more *efficient* free throw shooter of the two players. Most fantasy sports resources display these raw percentages only.

A closer look, however, reveals that Player B, though he is less efficient at the line, is actually the more desirable free throw shooter of the two players. The reason Player B is who we want to own is because, according to Ziguana, his FT% *z-score* (1.00) is higher than Player A's FT% z-score (0.50). How can a player who is less efficient at shooting free throws be more desirable to own? The answer is volume.

Continuing with the example, we can safely assume Player B *attempts* many more free throws than Player A attempts because Ziguana accounts for free throw volume in its z-score calculation. Based on this information, we know Player B is a more *active* player who draws more fouls and creates more scoring opportunities.

Managers desire players who are active *and* efficient and, using this example, Ziguana's FT% z-scores conveniently provide this information for us. Thanks to z-scores, we don't need to research how many times a player goes to the foul line. In fact, as strange as it might seem, we don't even need to know what a player's FT% is. All we need to know is a player's z-score, and the higher the z-score, the more valuable a player is in that category, plain and simple.

Another Ziguana feature I will tout is the site's ability to extract specific information from our Yahoo! and ESPN leagues. To make the most of the website, you need to upload your Yahoo! or ESPN league (requires less than a minute) and update your leagues' rosters (with a single click) as the season progresses. After that, Ziguana does the rest. When you run Player Ranks reports, you select which league you are currently researching and Ziguana loads the categories for that league. You can quickly adjust the categories to suit your specific search.

To add to the customization, Ziguana knows the ownership status of every player in your Yahoo! and ESPN leagues. *Your* players are visually identified by a gold star ★ beside the player's name. Players who are owned by *other managers* are identified by a red stop sign ●. Players in your league who are available *free agents* are identified by a check mark in a green circle ⊘.

The website's designers continuously improve the site and each improvement makes our analyses easier to perform. In fact, with the addition of a few new features, we'll be able to streamline some of the methods I describe in this manual. Ziguana team member Matt Honea states, "Ziguana's objective is simple: We want to help you build the best teams possible." I'll close this section with a listing of more Ziguana features that are currently available:

- Trade Evaluator – Analyze trade offers you receive or determine your own trade offers.

- My Teams Analysis – Analyze the players on your roster. When reviewing your team, you can select all your league's categories, or only a few. You can include data from the entire season, or from a selected length of time.

- My Leagues Analysis – Compare the strengths and weaknesses of your team to all the teams in your league.

- Head-to-Head Matchup Analysis – Projected results for H2H matches. Ziguana compares your team to your competitor's team and displays the results in a user-friendly format.

- Projected Statistics – Running projections for the remainder of the season based on mathematical regression models.

- Articles – Read insightful articles written by other Ziguana members and submit your own articles as well.

- Forums – Participate in friendly forums, ask for fantasy advice, and learn about site changes and improvements. You can even request a feature you want to see implemented on the site.

TIMEOUT 1

Players who are eligible to play multiple positions will provide your roster with added flexibility. Few things were more frustrating than the times I was forced to bench quality ball players – even when there was an open position that day – because I lacked a sufficient number of players who were eligible for that position. If I maintained players who, collectively, covered each of the five positions at least three times, I could usually avoid this benching headache.

The Multi-Position Player

Let's take some time and discuss Ziguana's z-scores and zTotals in more detail. In statistics, a z-score measures the difference from a data point to the average of the entire data set. Z-scores are expressed in units called standard deviations. A Ziguana z-score is a one-category measurement that tells us how many standard deviations a player's performance is from average for the selected category and time period. If you want to compare players using multiple categories, the Ziguana zTotal is the sum of all the z-scores for any given search.

In a Roto or H2H league, winning any category is just as valuable as winning another. Collecting the most steals is just as important as scoring the most points. Z-scores give us a standardized way to compare stats from different categories. You can't just add up a player's stats from chosen categories and compare your findings to another player's stats. Doing so would be like comparing apples and oranges. We can, however, add up a player's z-scores for the same categories and compare this zTotal to other players' zTotals for the same categories and time periods. In this way, z-stats standardize each statistical category and let us evaluate players across those categories.

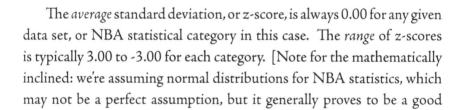

The *average* standard deviation, or z-score, is always 0.00 for any given data set, or NBA statistical category in this case. The *range* of z-scores is typically 3.00 to -3.00 for each category. [Note for the mathematically inclined: we're assuming normal distributions for NBA statistics, which may not be a perfect assumption, but it generally proves to be a good estimation for most statistical categories.]

In the screenshot on the following page, you'll notice LeBron James averaged 1.6 three-pointers (3PTM) per game during this past season and his z-score in this category was 1.00. Keeping in mind the average z-score for any given category is 0.00, it's clear that LeBron performed better than most NBA players in this category. Ziguana displays this positive z-score in green to indicate the player performed better than average in this category for the selected time period.

Your Custom Basketball zRanks:

Show player availability from my league: [Yahoo Public 179391 >]   Export to CSV

Current Season zRanks using YOUR categories:

zRank	Name	Team	Pos	GP	FG%	3PTM	FT%	REB	AST	TO	ST	BLK	PTS	zTotal
1	Chris Paul	NOR	PG	78	50.3 (0.64)	0.8 (-0.10)	86.8 (1.66)	5.5 (0.16)	11.0 (4.03)	3.0 (-1.81)	2.8 (4.74)	0.1 (-0.83)	22.8 (1.81)	10.30
2	LeBron James	CLE	SF	81	48.9 (0.52)	1.6 (1.00)	78.0 (0.31)	7.6 (0.98)	7.2 (2.17)	3.0 (-1.83)	1.7 (1.97)	1.1 (1.03)	28.4 (2.88)	9.03
3	Dwyane Wade	MIA	G	79	49.1 (0.62)	1.1 (0.30)	76.5 (-0.03)	5.0 (-0.05)	7.5 (2.27)	3.5 (-2.54)	2.2 (3.28)	1.3 (1.34)	30.2 (3.21)	8.42
4	Danny Granger	IND	GF	67	44.7 (-0.29)	2.7 (2.48)	87.8 (1.88)	5.1 (-0.03)	2.7 (-0.06)	2.5 (-1.08)	1.0 (0.23)	1.4 (1.58)	25.8 (2.37)	7.07
5	Kobe Bryant	LAL	SG	82	46.7 (0.09)	1.4 (0.74)	85.6 (1.50)	5.2 (0.03)	4.9 (1.00)	2.6 (-1.22)	1.5 (1.41)	0.5 (-0.24)	26.8 (2.57)	5.88
6	Dirk Nowitzki	DAL	PF	81	47.9 (0.32)	0.8 (-0.19)	89.0 (2.03)	8.4 (1.32)	2.4 (-0.20)	1.9 (-0.31)	0.8 (-0.45)	0.8 (0.36)	25.9 (2.38)	5.27
7	Jason Kidd	DAL	PG	81	41.6 (-0.36)	1.6 (0.98)	81.9 (-0.05)	6.2 (0.41)	8.7 (2.87)	2.3 (-0.82)	2.0 (2.70)	0.5 (-0.18)	9.0 (-0.83)	4.72

Now look at the *cumulative* zTotal for Chris Paul during this past season in this nine-category league. Paul's season zTotal was a league-leading 10.30. His performances in three-pointers (3PTM), turnovers (TO) and blocks (BLK), relative to all the other NBA players, were below average (indicated by red, negative z-scores), but his dominant performances in steals (ST) and assists (AST) were stellar.

One player's z-scores are relative to all other z-scores for the time period and categories you are analyzing. Z-scores are, in essence, a zero-sum-game scoring method. When one player's z-score improves (increases), another player's z-score worsens (decreases). Also, notice that Ziguana knows turnovers (TO) is a category in which fewer is better, and it scales the z-scores accordingly. Personal fouls (PF) are handled in the same fashion as turnovers.

Now let's focus on what's important for mangers when evaluating z-scores and zTotals. First, remember a z-score of 0.00 is not terrible, it's just average. About 67% of players will realize a z-score between 1.00 and -1.00 for any given category and time period. Roughly 95% of all players will score between 2.00 and -2.00 in a single category, and approximately 99% of all players will score between 3.00 and -3.00 for one category.

Practically speaking, a z-score of 1.00 for any given category should be considered a good score, a z-score of 2.00 is very good, and a score of 3.00 for one category is great. Any score beyond 3.00 is outstanding. Similarly, a z-score of -1.00 is considered to be weak, -2.00 is very poor, and anything below -3.00 is abysmal for that category and time period.

I've spent a lot of time discussing Ziguana and the site's statistical capabilities, but I haven't yet described how I utilized this data to evaluate players and manage my rosters. Please know I have repeated all my procedures in a condensed format in "The Playbook" section of this manual. If you follow my playbook, I'm confident you'll do well in your leagues. With that said, let's proceed to the next section where I'll discuss the heart and soul of how I managed my three fantasy leagues this past season.

2ND QUARTER

There is a dynamic element to sports, and hopefully some predictability in "hot" and "cold" trends. The question is… How can we best predict future performance from historical data? The results from my research on momentum or the "hot hand" have been shaky and inconclusive. The point is projections are tricky. We at Ziguana are constantly developing ways to determine statistically significant means of predicting future performance.

Cassidy Morris, Ziguana.com

The screenshot shown on the next page shows a portion of the spreadsheet I developed during the fantasy NBA season. The spreadsheet displays most of the pertinent information I recorded to help me evaluate my players and monitor some of my leagues' activities. I spent months developing it and trying to make it as simple and unassuming as possible. As the season unfolded, I continuously improved the spreadsheet, and I'll surely improve it more during the offseason and throughout future seasons as well.

In the appendix of this playbook, I offer detailed explanations of every cell value in the spreadsheet. I also include all the formula and conditional formatting cell information. If you're uncomfortable creating spreadsheets, or if you simply want to save time, you can download the spreadsheet at the following website: www.awinningplaybook.com.

Although I'll reserve the spreadsheet detail for the book's appendix, here I'll explain the big picture ideas behind this useful tool. To begin, every time I added a player to one of my rosters, I adjusted the 'Player' section to reflect the change. Also, once every week during the season, I recorded each player's cumulative zTotals for all my league's categories for three time periods using Ziguana: full season, the last 45 days, and the last ten days.

I used this historical data in my attempt to predict the future performance levels of each of my players. I assigned weighted averages to each of the zTotals from the three time frames. I then created an expected value formula to calculate my performance predictor, which I refer to as the "Expected zTotal."

TEAM NAME

									SEASON zTotal	45 DAY zTotal	10 DAY zTotal	EXPECTED zTotal
									1.31	1.92	2.52	2.04

									SELECTED CATEGORIES			EXPECTED
ACQ. DATE	DAYS OWNED	PG	SG	SF	PF	C	R/W	PLAYER	SEASON zTotal	45 DAY zTotal	10 DAY zTotal	zTotal
29-Oct	123	1					*	PLAYER 01*	9.03	9.44	9.86	9.53
29-Oct	123				1	1	*	PLAYER 02*	4.72	3.36	3.42	3.51
1-Feb	28			1	1		•	PLAYER 03	(2.21)	1.90	5.87	2.68
29-Oct	123		1		1		•	PLAYER 04	2.10	2.71	2.59	2.61
29-Oct	123			1			•	PLAYER 06	3.62	2.42	2.08	2.44
14-Dec	77					1	•	PLAYER 05	0.52	2.68	2.30	2.35
20-Feb	9					1	•	PLAYER 07	(1.20)	1.15	1.95	1.16
14-Nov	107		1				•	PLAYER 08	(0.72)	0.46	2.51	0.96
15-Jan	45		1				•	PLAYER 09	1.01	1.06	0.41	0.86
14-Jan	46				1		•	PLAYER 10	0.80	0.23	1.61	0.70
1-Mar	0					1	•	PLAYER 11	0.06	0.52	0.56	0.49
1-Dec	90		1				•	PLAYER 12	(0.79)	0.29	1.26	0.47
14-Feb	15		1				•	PLAYER 13	0.10	(1.30)	(1.61)	(1.25)

CAN'T CUT*
TRADE OFFERED
TRADE PENDING
ATTENTION REQUIRED

2	PG
7	SG
4	SF
4	PF
4	C

TIER 1 MINIMUM zTOTAL	2.50
TIER 2 MINIMUM zTOTAL	0.50
TIER 3 MINIMUM zTOTAL	(1.00)

Predicting future athletic performances is the crux of fantasy sports and there are many variables to consider. Should we include last season's data in our formula? Should we emphasize more recent performances? How important is the strength of a player's team? Should we adjust our formula as the season unfolds? I don't claim to have perfect answers to these questions. If we knew all the answers, then fantasy sports management would not be the challenge it is. I'm simply offering flexible and effective tools we can use to help predict future performances and to help us win our fantasy leagues.

As I mentioned, once every week during the season, I recorded each player's Season zTotal, his 45-day zTotal and his 10-day zTotal for all the categories my leagues scored. The quickest way to obtain this information is by using Ziguana's 'My Teams' page. Collecting and recording these zTotals requires less than five minutes per team. These cells are conditionally formatted to change color to gold, green, grey or red based on the value you enter in the cells.

Every league will have its star performers, its average performers, and its weaker performers. Gold cells visually represent the highest performance level, or the "gold standard" of performances. I refer to this highest performance level as the 'Tier 1' level. Green cells represent good, or respectable, performance levels, and I refer to this performance level as the 'Tier 2' level. Grey cells indicate mediocre, or 'Tier 3', performance levels. Red cells indicate the player's performance level for the selected time period is very poor. By viewing these colorfully illustrated data from left to right for each player, it's easy to see the performance trends and to determine who is performing well and who is underperforming.

It's important to accurately determine the minimum zTotals that should define the three performance levels in your league. The appropriate zTotals depend on three variables: the size of your league, the number of players on each team, and the number of categories your league scores. In the appendix, I provide detailed guidelines that explain how to determine appropriate tier-defining zTotals for any given league.

The Expected zTotal is my way of valuing any player in the NBA, and it's my attempt to predict a player's future performance level. I experimented with many different Expected zTotal formulas during the season, and the formula I ultimately settled on is the following:

Season zTotal	= 10% of the Expected zTotal
45-day zTotal	= 60% of the Expected zTotal
10-day zTotal	= 30% of the Expected zTotal

It's not critical to use "perfect" performance predictors. In fact, it's impossible to do so because we can never predict the future with 100% certainty. I encourage you to experiment with different Expected zTotal formulas to find the formula you believe works best. What is most important is for you to continuously monitor your player's performances and to compare these performances to their peers' performances when you make your managerial decisions. Regardless of the performance predictor formula you choose, Ziguana allows us to effectively make these comparisons using the categories and scoring formats of our specific leagues.

As the season progressed, my goal was to acquire and retain players whose Expected zTotal for all of my league's categories met the Tier 1 (gold) or Tier 2 (green) performance standards. Occasionally I made exceptions to this gold-or-green rule, but by and large I adhered to my Tier 1 or Tier 2 Expected zTotal performance standards. If you're able to continuously maintain a full roster of players who perform at the Tier 2 (green) level or better, your team will be very difficult to beat.

When researching player performances, which I'll discuss in the next section, I regularly scouted for free agents who had stellar z-scores and zTotals, and I continuously replaced my underperforming players with these fresh studs. After weeks of routinely adding proven players to my teams and dropping my weakest players, my add/drop decisions became increasingly more difficult.

My decisions were harder to make, not because I grew fond of the players on my team, but because the players on my teams were all superlative. The day I targeted a powerful free agent to add to my roster, but I elected to do nothing because this fine player would not have significantly improved my team, I achieved a milestone. At that point, for the first time during the season, I realized my team was as good as it could be, and I stopped searching for free agents altogether...for an entire week!

HALFTIME

Owners should maintain a balance of positional players on their rosters, especially in rotisserie leagues. It doesn't hurt to have an abundance of players who are eligible to play any given position, but it certainly hurts to have too few. By maintaining a minimum of at least three players who were eligible to play each position, I was able to accumulate the maximum number of games allowed in all my leagues for every position.

In this 'Halftime' section, we definitely won't be taking a break (but feel free to grab a towel and some Gatorade). Here I'll describe the Ziguana zRanks reports that I ran virtually non-stop for five months. After countless hours of analyzing reports from different time lengths, and after monitoring the performance levels of practically every valuable free agent in my leagues, I can offer a systematic research process that is easily repeatable and highly effective. I spent most of the season scouting for free agents, and I'll spend a good bit of time in this section describing how I conducted my searches.

Before you search for free agents, be sure your league rosters are up to date at Ziguana by clicking 'Update Rosters' on your 'My Leagues' page. Then go to 'Player Ranks' and generate a report using all the categories in your league. For most basketball analyses on Ziguana, the reports default to using per-game averages as opposed to using season totals. [On the baseball side of the website, the reports default to using totals, as is traditional when discussing baseball statistics.]

Next, check the box labeled 'Return Only Free Agents.' [As stated earlier, you'll need a Yahoo! or ESPN league to utilize this feature.] After you update the stats, you'll see the top free agents based on the full season's data. During the first six weeks of the season, use data from the full season for your analysis. After the season is at least six weeks in, only consider stats from the last 45 days.

Next, open a new window in your browser (or a new tab if your browser supports tabbed browsing). In this new window or tab, run the same report, except this time use stats from only the last ten days. Now you're viewing the free agents in your league who are currently "hot" and who are playing at a relatively high level. Compare both reports. Focus your search on players who have been performing extremely well during the last ten days and who have played at least respectably well during the last 45 days. In other words, your top prospects should be consistently solid players who are also currently playing at a very high level. There are three types of players you'll see as you view these free agent reports, and I'll describe each type.

1. High 45-day zTotal, Low 10-day zTotal: If you locate a free agent who has been playing very well, on average, for the past six weeks, but his performance level has dropped significantly during the past ten days, my suggestion is to *not* acquire this player and to continue your search for a different free agent. We know this free agent has the potential to play at a high level, and one can argue that it might be advantageous to add such a player to your roster now, before he raises his game again and other managers take notice. Later I'll offer my opinions about "buying low," but for now I'll just say that I prefer to acquire players who we know can play well *and* who are playing well *now*.

2. Low 45-day zTotal, High 10-day zTotal: If you locate a free agent who has been playing decently for the past six weeks (good, but not great zTotal), but his stats have improved significantly during the past ten days, my suggestion is to *consider* acquiring this player, but to continue your search. This player could continue his hot streak, but he could also be experiencing a fluky span of fine play. It's also possible this player has been playing more minutes recently, thus allowing him more court time to improve his stats.

When making our final decisions in cases like this, it helps to determine whether recent improvements in a player's stats are a result of *better* play, or *more* play, or both. To aid with this determination, Ziguana conveniently offers 'minutes played' as a category statistic you can include in any customized report. You can also have Ziguana analyze players' stats by using *cumulative* totals for a given time period instead of per-game averages.

3. High 45-day zTotal, High 10-day zTotal: Bingo! When you discover a free agent who has been playing very well for the past six weeks, *and* he has maintained or exceeded his high performance level during the past ten days, my suggestion is to research this player further and to seriously consider adding him to your team. If the sportswriters you rely on confirm this player

is healthy, you will likely want to pull the trigger right away and add him to your roster.

Sometimes you'll discover free agents with such high 45-day and 10-day zTotals that you'll wonder how the player can be available in your league. These are the times that can generate the fantasy management excitement we've all experienced. When you discover such a player, and if you also have a player on your roster who has been seriously underperforming, the excitement is even greater as you rush to your league's website to execute the add/drop transaction immediately.

Of course there are no guarantees that any player we acquire will perform well. In the end, we simply attempt to increase our chances of success by acquiring ball players who consistently outperform their peers. After we acquire these players, we continuously monitor their performances (with the use of the spreadsheet described earlier) and we make future add/drop decisions based on new and constantly changing data.

Using Ziguana's player ranking reports, we can perform these types of free agent searches for any number of selected categories. For example, we can search for a free agent who is a great defensive player by selecting the categories in your league that are defense related (e.g. steals, block and rebounds). Keep in mind that cumulative zTotals are typically lower when analyzing fewer categories. We can also search for free agents who excel at *one* specific category (a specialty player), but I'll reserve my discussion of this special search for 'The Buzzer Beater' section of this book.

When you have selected a suitable free agent, you must decide which player you will drop from your roster. Hopefully this will be a *difficult* decision for you because this would indicate that all your players are relatively strong. Use my spreadsheet and my Expected zTotal value to help with your decisions. As tempting as it can be, try not to penalize players on your roster whose 10-day z-scores are low, especially if their 45-day zTotals remain solid. Such a drop in performance could be short lived and you might regret releasing such players prematurely.

On the other hand, you should highly value players on your rosters whose 10-day zTotals are strong, even if their 45-day zTotals are less impressive. These players have either recently improved their game significantly, or they are playing more minutes, or both. Regardless of the reasons for their increased production, it's good news for you, and you should think twice before releasing such a player.

Here's a tip managers should implement when making add/drop transactions. Just before you add/drop, compare the upcoming schedule of the player you intend to drop to the schedule of the player you plan to acquire. [To compare upcoming schedules at Yahoo! Fantasy Sports, simply click on 'Opponents' when you're viewing the potential transaction.] Confirm the team of the player you will add is scheduled to play at least as many games in the coming week as the team of the player you will drop.

One of your goals is to maximize your number of games played. If you determine the free agent's team is scheduled to play significantly fewer games in the next several days than the team of the player you want to drop is scheduled to play, consider waiting before you make the transaction. By carefully reviewing each team's immediate schedule, you can determine the ideal day to execute your add/drop transaction, thus maximizing your total games played.

Keep in mind that delaying add/drop decisions involves risk. The longer you wait to add your newly found prize, the more time you allow your competing owners to discover this player and to acquire him before you do. Assuming an immediate add/drop isn't critically necessary, my advice is to wait and make the transaction when you will realize the maximum number of games played in the coming week. However, if other managers in your league are snagging your target players before you acquire these players, then you should make your future add/drop transactions immediately in that league.

How often should we search for quality players to add to our teams? The answer is purely subjective, and I'll offer my advice. In short, unless all the players on our roster are superlative, I believe we should be searching for valuable free agents once or twice per week. In addition, we should allow at least three or four days between searches. Searching too seldom is a trait of an inactive manager, but searching too frequently can be counterproductive and a waste of valuable time.

Some might argue that conducting free agent searches every week is too frequent, and that adding new players this often qualifies as a "flavor of the day" strategy. To these antagonists, I would simply remind them that *minutes played* is a crucial variable in the NBA. Given the dynamic nature of coaches, owners and the players themselves, quality ball players who typically ride the pine can often be thrown into action on very short notice. By researching free agents every week, we can sometimes detect these increases in minutes played (by a sudden and significant increase in z-scores) before we learn about the roster changes from NBA news sources.

How many transactions should fantasy managers make each season? Inactive managers who make only a few transactions during the entire NBA season could suffer from stale and ineffective teams. At the opposite extreme, managers who trade multiple times each week are likely not allowing sufficient time for their players to prove themselves. According to the results of a study conducted by Patrick of GiveMeTheRock.com, fifteen seasonal transactions might be the point of diminishing returns.

Bret Morris of Ziguana.com makes a good point by stating, "Too often managers will jump at the hot player and jump ship when a player hits a rough patch." I was guilty of this more than once during my rookie season. I predict that next season I'll complete between 25 and 40 transactions in my roto leagues, and I'll execute about the same number of transactions during the regular season of my H2H leagues. For reasons I'll explain soon, H2H leagues require many additional transactions during the playoffs.

3RD QUARTER

Reply to trade offers. All you have to say is 'No' if you're not interested. It's simple, it takes about 11 seconds, and a robot would do it almost instantaneously, but it remains absolutely critical. Letting a trade offer expire because you didn't click "Accept" or "Decline" is one of the biggest sins in fantasy basketball. If you do reject a trade, make a counter offer. This is not mandatory, but it is highly recommended for the sake of everyone's enjoyment.

Matt Stroup, Rotoworld

When will the trade offer be reviewed?

When considering trades, I was fairly conservative during my rookie year. I did execute some effective trades during the season, and I was fortunate the trades turned out as well as they did. In this section I'll offer a Ziguana-based trade procedure that can potentially catch other Yahoo! managers off guard. I'll also offer some important psychological trade considerations that can arise in specific situations. I'll begin with the Ziguana procedure.

First, rank all players using the categories of your choice. Next, check the box labeled 'Show Yahoo! Ranks' and update the stats. When the list refreshes, sort by the difference column by clicking 'Diff.' The players will now be ranked, in descending order, by the numbers in this column, and the numbers at the top of the column will be green (positive).

According to the difference between Ziguana's season zRanks and Yahoo!'s season yRanks, you are now viewing the players in your league who might be underrated by other managers. When considering the categories you selected, these underrated players are performing at a level higher than the Yahoo! season ranks might lead your competing managers to believe. The underrated players who are already owned in your league are players you might consider targeting in a trade offer.

Note: The above section is *not* intended as a slight against the Yahoo! ranks. Yahoo! ranks are extremely convenient and I use them all the time. Ziguana's zRanks, however, consider z-scores for the categories and time lengths of your choosing. In essence, Ziguana allows us to create truly customized player rankings for our specific league, and we can compare these custom rankings to the rankings provided by other sources.

Back to the procedure... Now you must determine which of your players you will offer in your trade. In a separate browsing tab, run the same report, and then filter for *your team only*. Display the Yahoo! ranks and again sort by the 'Diff' column. The players at the *bottom* of this report, as indicated by the relatively large (red) negative numbers, are the players on your roster who might be *overrated* by your fellow Yahoo! managers. An ideal trade would consist of all the following criteria:

1. Your target acquisition has a high (positive) green number in the 'Diff' column;

2. Your player has a low (negative) red number in the 'Diff' column;

3. The zRank of your target acquisition is lower (better) than the zRank of your player; and

4. The yRank of your target acquisition is higher (worse) than your player's yRank.

If all these conditions exist, you might consider making a trade offer for the target players. If the manager who owns your target acquisition relies primarily on the Yahoo! season ranks to determine his player's values, you might be able to execute an advantageous trade.

Here is some trade advice specific to managers in H2H leagues. Near the end of the regular season, assuming *you have already secured a spot in the playoffs*, review the rosters of all the owners who are still fighting to make the playoffs. Look for strong players on these teams who are currently *injured*, but who should return to action soon. If you locate such a player, consider making a lopsided trade offer for this injured star. Take advantage of the vulnerable owner and offer him considerably less value than you would offer him under normal circumstances.

Consider the other manager's position. He's stuck with an injured player when he desperately needs all his guys on the court. If this manager cares about how he finishes the season, he'll consider *any* offer that might help improve his chances of securing a playoff spot...wouldn't you? That said, be careful about trading for an injured player *if you are fighting for a first round bye*. The last thing you want is to acquire an injured stud who cannot help you for a week or two, only to fall to third place in the overall standings (thus missing a first-round bye) while your soon-to-be-activated player sits on your bench.

When deciding which players to target for a potential trade offer, if all other variables are equal, prioritize players from teams that are *below* you in the overall league standings. Owners who are leading your league might be very conservative, or risk averse, when analyzing trade offers. These managers don't want to lose face by accepting a potentially advantageous offer, only to feel burned later when the trade works against them. Consequently, you might need to offer these league-leading managers more value with your trades.

Teams at the bottom of your league, however, might be willing to assume more risk when evaluating your trade offers. These managers might think they have little to lose by accepting incoming offers. This can be especially true if you are deep into the season or if you're approaching the trade deadline. Regardless of *your* overall position, any manager who feels desperate to improve his league standing should seriously consider *any* trade offer from *any* competing owner. Of course, we shouldn't insult anyone when offering trades, but the psychological aspects of league position as it relates to trade offers should not be ignored.

TIMEOUT 2

Fantasy MVP Award 2008-09:

I'm going with Chris Paul. CP3 led the league in steals and assists for the second straight year, becoming the first player in history to do so, and he posted one monster line after another in 78 games. Paul's owners likely got off to a slow start this season with all the early three-game weeks, but the Hornets' tremendous second-half schedule helped make up for the slow start in the fantasy version of the tortoise vs. the hare. If you traded Paul prior -to the break, it was likely a devastating error. For those of you in support of LeBron or Dwyane Wade winning this award, they were very close.

Steve Alexander, Rotoworld

I'll call a timeout to look at a fantasy sports analogy: that of trading equity stocks on the open exchanges and managing one's financial portfolio. Many fantasy managers and stock traders follow the popular adage, "*buy low, sell high.*" This is surely an effective strategy...when you can achieve it. I will contend, however, that it might be more advantageous when considering roster changes to "*buy high and hold*" players until the players' performance levels begin to deteriorate.

Some stock operators believe we should search for stocks that are making new price highs and breaking out of consolidation (sideways moving) areas. Similarly, I choose to "purchase" players who have already "broken out" into high performance levels and who have already proven to be valuable assets. Then, if the player proves me wrong, I release the player and I replace him with another "stock" whose "purchase price" is relatively high.

When speculating in the stock market, "investors" purchase equities they believe are undervalued (much like you can do by comparing zRanks to yRanks). After they purchase, they often wait for the stock price to increase to a level they believe the stock should be worth. These types of investors purchase for the long run, and they might not be concerned if the stock's price drops significantly after they purchase it.

Stock "traders", on the other hand, often purchase stocks that have *already* risen to a high price. These speculators are not concerned about purchasing a stock others believe to be overvalued. Instead, traders often pay a premium for their shares and, after they purchase, they typically won't allow the price to drop too much before they sell the stock and cut their losses. These operators believe it is less risky to purchase a high quality stock at a relatively high price because this proven stock has demonstrated that its upward trend has a good chance of continuing.

Likewise, in fantasy sports, managers sometimes behave like investors and we sometimes behave like traders. Acquiring a rookie who has never played an NBA game (or a major league game in the case of baseball) is one example of investing in fantasy sports. Another example of fantasy sports investing is acquiring a perennial star when the player is injured or experiencing a poor season. After acquiring such a player, managers then hope for the player to return to his typical domineering form.

An example of fantasy sports trading would be acquiring a top-performing free agent, and then releasing the player after a short period of time if his stats drop significantly. Another example of trading in the fantasy sports realm might be offering another manager "too much" value for a player who is currently outperforming his peers. Because I'm relatively active in both fantasy sports management and in the stock market, and because I tend to purchase only proven assets, and sometimes at premium prices, it's probably no surprise that I consider myself to be more like a trader than an investor when operating in both realms.

Earlier I stated it might be best to avoid the buy low, sell high strategy. One of the risks involved with this strategy is the risk of *time*. To buy low, we first spend time researching players who we think will soon break out. After we acquire the player, we spend time waiting for this player to raise his game. If the player *does* break out, we spend time determining when to trade him, and to which manager we should trade him. Then we spend time waiting for the other owner to research our player and to analyze our offer. If our trade offer is accepted, we might continue monitoring the player to see if his performance level drops as we anticipated it would. If, after we trade the player away, the player continues to play at a high level, we spend time berating ourselves for our mistake. Every minute we spend performing these tasks represents expensive time-related opportunity costs, and time is too valuable to waste.

Buying *high*, on the other hand, arguably carries less risk. First, we have an efficient method for locating valuable free agents. When we spot a great free agent, and when we have a player on our roster who should be dropped, our add/drop decision is easy and we waste no time picking up the new player. Within minutes, we make our roster change and our "portfolio" instantly improves. If the player proves us wrong later, we waste little time in dropping him.

I'll finish the buy high argument by going so far as to say it can be risky to *not* acquire a powerful free agent when we believe he would be a good fit for our team. By passing on this free agent, we allow our competing managers the opportunity to claim this player and to enjoy the benefits of ownership that *we* should have enjoyed. In most cases, I'd rather pick up the player and drop him later, than to allow another owner the chance of acquiring the potential stud.

4TH QUARTER

Inactive managers who make only a few transactions during the entire NBA season could suffer from stale and ineffective teams. Conversely, managers who trade multiple times each week are likely not allowing sufficient time for their players to prove themselves. It's important for managers to remain active throughout the season, but it's also important for us to sit back and observe, and to stop ourselves from making rash decisions.

We need to address the importance of evaluating player performances in leagues of different sizes. By league *size*, I'm referring to the number of owned players within the league. A player whose relative performance level qualifies as "great" in a large league might only qualify as "fair" in a smaller league. It's critical we maintain proper perspective when switching between these different sized leagues. We must set objective, observable parameters to distinguish these differences and, not surprisingly, it's all about z-scores.

The fewer the teams in a fantasy league, the fewer NBA players will hold enough value to be worthy of ownership in the league. When fewer players are valuable, our minimum desired performance levels for players we want to acquire naturally increases. If you follow the procedure I outline in the appendix, you should have no trouble objectively defining performance levels that are "great", "good", "fair" and "poor" in your specific leagues.

You'll see in the appendix that in my *twelve*-team, nine-category league, I defined the tiers with the following minimum zTotals: 2.50 (Tier 1), 0.50 (Tier 2) and -1.00 (Tier 3). By comparison, in my *eight*-team, nine-category league, I defined the tiers thusly: 3.50 (Tier 1), 1.50 (Tier 2) and 0.25 (Tier 3). In the upcoming 2009-10 season, if I join leagues that have the same number of teams and score the same categories, I will use the same tier-defining zTotals.

Now that we know our required zTotal expectations will be different in leagues of different sizes, we can direct our focus on the free agent search. When you run your player ranks reports in Ziguana, always consider which league you are searching in (assuming you join leagues of different sizes). The zTotals of the top-ranking free agents in your smaller leagues will tend to be higher than the zTotals of the free agents in your larger leagues. As you perform these searches, try to keep a specific zTotal value in mind.

For instance, if you join a twelve-team, nine category league, you might use a "gold standard" zTotal threshold of 2.50, as I discussed earlier. If the top free agent in this league scores a 45-day zTotal of 2.50 or higher, and his 10-day zTotal is even greater, this free agent should be worthy of further consideration. On the other hand, if you join an eight-team, nine category league, you should demand higher zTotals from the free agents you consider. In such a league, you might elect to use 3.50 as your minimum desired zTotal. The exact target you choose is not as important as simply remembering to use different zTotal targets when searching within different-sized leagues.

THE BUZZER BEATER

At the beginning of a season, punting categories can be considered a strategy. As the season progresses, ditching categories becomes more of a technique that can be fine tuned as we observe what's happening in our leagues. It's liberating when we know we can safely ignore one or more categories. Of course losing a category is not good, but if we determine our efforts to gain ground in the category will be futile, then we can discount the category altogether and focus our attention elsewhere.

As the season unfolded and my category scores in each league changed, my free agent searches evolved. In one roto league, I gained such a big *lead* in assists that later in the season I removed assists altogether from my free agent search considerations. In my other roto league, I became so far *behind* in turnovers that about halfway through the season I ignored turnovers during my searches. In one league, my standings in both 3PTM and A/T were in the middle of the pack, but the spreads between the teams ahead of me and the teams behind me in both categories were so great that I removed these categories from further consideration.

Side note: Regarding our league's category standings, Yahoo!'s 'Distance from Competition' feature in the overall standings section is extremely useful. Users can sort league standings by cumulative scores or by one specific category. Doing so allows us to easily determine, both visually and numerically, in which categories we need to improve.

At the beginning of a season, punting categories can be considered a *strategy*, but as the season progresses, ditching categories becomes more of a *technique* that can be fine tuned as we observe what's happening in our leagues. It's liberating when we know with absolute confidence that we can safely ignore one or more categories. Of course losing a category is not good, but if we determine our efforts to gain ground in the category will likely be wasted, then we can discount the category altogether and focus our attention elsewhere.

In this regard, we can differentiate between "ditching" and "ignoring" categories. We can choose to ditch categories at any time during the season, but we should only do so if we are certain we will finish poorly in the category, regardless of our efforts (as in the turnovers example above). Alternatively, we can ignore categories anytime we believe our efforts to improve our category score will be futile. This can occur in categories we are losing *or* winning. To state the concept differently, we always ignore categories we are ditching, but we don't necessarily ditch the categories we are ignoring. Regardless of whether we are ditching or ignoring categories, the fact that we can subsequently focus our attention on narrower searches is extremely beneficial.

When you make a decision to remove categories from further consideration, it's important to go all out. After I stopped considering specific categories in my free agent searches, I repeatedly reminded myself to do so *completely and for the remainder of the season* unless some unforeseeable change occurred. In addition, I also removed the same categories from my player evaluations and my zTotal entries on the spreadsheet. I quickly learned the benefits from these narrower free agent searches. As the season progressed, I found myself searching for free agents who excelled in fewer and fewer categories, and during the final weeks of the regular season, I was searching mainly for "category kings."

Category kings, or specialty players, are players who excel in at least one category and who, ideally, will not hurt you too much in other categories. As the end of the season draws near, one point here or there in the league standings can make the difference between winning and finishing in second place. In a category race that's very close, managers need to acquire players who excel in this specific category. There are several methods owners can use to search for category kings and I prefer to use Ziguana's player ranks reports.

On the Player Ranks screen, sort all the players in your league using *all your league categories*. Filter for free agents only and restrict your search to the past 45 days. In the header row, click on the category name in which you need help and Ziguana will sort all the players – by z-score – by this category. Open a second browsing tab and create the same report, but this time restrict your search to only the last ten days. Again, sort the players by your category. After reviewing the two reports, you will likely see a few suitable players who will help you improve in your selected category.

When searching for category kings, it's helpful to view *all* your league's categories instead of running the report for one category only. Doing so provides us with a broader perspective of the players' overall skills. After you narrow your selection to just a few players, by viewing the z-scores of all your league's categories, you can determine in which other categories your candidates might hurt your team. This information helps make your decision easier.

The last step, of course, is to select a player from your roster whom you will drop to make room for your target specialty player. To do so, run the same reports, but this time filter for your team only before you sort by the specific category. By comparing your players across all your league's categories for the past 45 days and for the past ten days, and by focusing on the category in which you need help, you will be able to determine which player, or players, you should target for release. One final comment about acquiring specialty players: if you require help in a *ratio* category, you should consider adding *at least two* specialty players instead of adding only one. By acquiring more players to your team who specialize in a ratio category, you will gain more volume, and as previously discussed, volume is an important factor in these types of categories.

THE PLAYOFFS

Head-to-head playoffs in a daily lineup league is a bizarre creature, a place where many of the established fantasy protocols vanish, leaving you with nothing to do but accrue as many games as possible to advance to the next round. Basically, you're going to want to set your alarm for 6 am to add/drop approximately two to three new players every day, no matter what. You can go back to sleep afterward.

Matt Stroup, Rotoworld

THE ACCUMULATION STRATEGY

I had a blast during my H2H playoffs. As my rotisserie league activities were calmly winding down, my head-to-head playoffs suddenly picked me up again and whirled me back into action…after my first round bye, that is. Just before the quarterfinals started, I read a timely article by Matt Stroup of Rotoworld about fantasy basketball playoffs in daily leagues. Matt is right; daily H2H playoffs are, indeed, strange and wild. The article clued me in to what my objective should be: accumulate as many games played as possible every day, without dropping my most valuable players. This can be called the "accumulation strategy."

The article opened my eyes and I began to think of ways to efficiently determine which NBA teams were scheduled to play the following day, which players from those teams were free agents in my league and, finally, which of those available free agents were currently playing well or playing lots of minutes. It was a lot to consider, and it took me a while to figure it out, but it eventually became clear. I'll take you through my approach, step by step. Keep in my mind the procedure considers Yahoo! leagues only. Also, Ziguana will likely offer additional features soon that will help us improve this procedure.

First, at the Yahoo! league page, go to the 'Players List' search page. Yahoo! conveniently includes a column that displays the next day's opponents for each listed player. If there is no listed opponent beside a player's name, then that player's team is not scheduled to play the next day. At the stats pull-down field, select 'Ranks.' First sort the players by their 'Season' ranks, then sort them by their 'Last Month' ranks, and finally sort them by their 'Last Week' ranks.

After each sort, identify every player listed on the first screen (the top 25) who is scheduled to play the following day and who is not injured (we all know the injury icon!). Click the blue flag in the 'Action' column next to these players' names, which places them on your Yahoo! watchlist. After you complete all three sorts and you identify and flag all the players who are scheduled to play the next day, your watchlist will be very long. In fact, on several occasions I maximized the number of players Yahoo! allows us to add to our watchlist.

Next, go to Ziguana and click 'Update Rosters' for your league. Go to the Ziguana Player Ranks page and select all the categories from your league, except turnovers. To repeat, exclude turnovers from your search. Later in this section I'll explain why we should exclude turnovers for our free agent searches during the playoffs. My leagues did not count personal fouls against us. Personal fouls is another "negative" category, and you should exclude *any* negative category from these searches. Run the report and then filter for the 'My Watchlist' players.

Ziguana monitors our Yahoo! watch lists the same way it monitors our team rosters. Ziguana allows us to filter our searches for our watchlist players, and we can compare these watchlist players using the categories and time frames we choose. Because watchlists change immediately (as opposed to roster changes which occur the next day), Ziguana is able to update any changes to our watchlists immediately as well. All you need to do is click 'Update Rosters' after each of your Yahoo! watchlist changes.

Back to the procedure… Restrict your search to the last ten days only, and there you have it… The players at the top of this list are available free agents in your league who are scheduled to play the very next day and who are currently playing well… *hopefully*. I say "hopefully" because it's possible the highest cumulative zTotal you will see is not amazingly high, but this is acceptable when you're only searching for warm bodies to throw on the court for one game.

After you target the best available free agents, you'll need to decide whether it's worth dropping some of your current players *who are not scheduled to play the next day* to acquire some of these players who *are scheduled to play*. As you might have guessed, I took this process to the extreme in my effort to accumulate as many of the countable categories as possible (e.g. 3PTM, REB, AST, ST, BLK and PTS). During the fantasy finals, I averaged almost four add/drops per day, and one day I dropped *eight* of my thirteen players who were not scheduled to play the next day so I could add eight free agents who were scheduled to play!

As the real NBA regular season nears the end, some teams are fighting to make the playoffs, and some teams that clinched a playoff spot are fighting for home court advantage. Players on these teams will typically continue to play hard to earn more wins for their teams. Other players, however, take days off during this time. Some might not play in their effort to rest and prepare for the playoffs. Others might not play because they're shutting down for the year and they want to prevent personal injury. The fantasy impact on these occurrences is real and sometimes significant, and you can use this information to your advantage by tweaking your free agent searches just a bit during your playoffs.

When your ten-day search is complete, open a second browsing tab and generate the same watchlist report as described above, but restrict your search to just the past five, or four, or three, or even *two* days. Compare this new report with the ten-day report and look for positive trends. This double-window search is not unlike the free agent search and analysis I described earlier using 45-day and 10-day time periods.

When searching for valuable players during the main part of the regular season, we're effectively looking for players who we believe will perform well for extended periods of time. To contrast, when searching for valuable players during the *playoffs*, we're really just looking for players who we believe will add stats for us *tomorrow*. We don't need tons of historical data to determine this. We just need to know what's happening *now*. Obviously if you locate a high quality free agent during the playoffs, you should consider keeping him for more than a day or two, regardless of his team's immediate schedule.

Always remember to check a player's latest news before you add him to your team, even if you intend to drop him after just one day. There are many things that limit players' minutes, including quarrels with the coach and suspensions from the league, which can cause your newly found player to be planted squarely on the bench, regardless of his recent performances.

It's time to discuss the negative categories. As stated earlier, the TO category is considered a negative category because each turnover players commit count against their owners. Turnovers is a category that's common to many leagues. I gave much consideration to turnovers during the year. Turnovers were scored in one of my rotisserie leagues, and they were a factor in my other rotisserie league (the league scored the assist-to-turnover ratio). Turnovers were also counted against us in my H2H league.

Note: Personal fouls (PF) is another negative category, but I did not join a league that considered personal fouls. In the discussion that follows, you can assume that personal fouls can be managed the same as turnovers. Personal fouls should not, however, be directly compared to the assist-to-turnover (A/T) category because PF is a countable stat, whereas A/T is a ratio stat.

I was fortunate that, in both of my rotisserie leagues, my turnover headaches remedied themselves. In my roto league that scored turnovers directly, I drafted a stud player who also turned the ball over several times each game. As the season unfolded, it became clear that I would not gain ground in the turnover category with this player on my roster. Instead of trading the player away, I simply punted the turnover category altogether and my future add/drop decisions became much easier as a result.

In my roto league that scored A/T, I closely monitored the category week after week. I discovered that midway through the season, when all the teams had accrued significant volume in assists and turnovers, my A/T quotient and ranking didn't change much, regardless of which players I had on my roster. Maybe this will occur again next season, maybe it won't. It *did* occur this year and, because I noticed it, I eventually disregarded this category with confidence, again simplifying my future add/drop decisions.

In my H2H league, I ended the regular season with a miserable winning percentage in TOs. That was not intentional; it just happened. Fortunately in H2H leagues, we can afford to bomb a category or two and still win our matches and finish well for the season. In the playoffs, however, because we plan to accumulate as many games played as possible, we will naturally amass more turnovers than our opponent. We almost *plan* to lose this category based on the nature of our strategy, so it must be correct to disregard turnovers when considering which players to acquire during the playoffs.

With that said, an intriguing concept arises if our opponent also utilizes the same accumulation strategy during the playoffs. If this occurs, the turnover category becomes winnable for us again, thus leading us to reconsider turnovers in our search for free agents. I'm certain, however, I will *not* consider turnovers next year during my daily league playoffs (assuming I make the playoffs), *especially at the beginning of each match*. Instead, I will aggressively plan to accumulate more games played than my opponents and, to be consistent with this strategy, I will initially ignore turnovers. Days into the match, if I determine that turnovers are winnable, I will act accordingly, assuming I will not sacrifice any other countable categories.

After I determined it was in my best interest to exclude turnovers from my player research during the playoffs, I considered excluding *all ratio categories* as well. As I threw bodies on the court each night, a funny visual kept recurring in my mind. Apparently what I was targeting, whether consciously or not, was a barrage of ball players haphazardly running around the court and whirling basketballs everywhere for 48 minutes! Given this chaotic (and hilarious) scene, my players would certainly accumulate many of the countable stats.

Although this might be an exaggeration of the accumulation strategy, it leads to a worthy consideration: during the playoffs, if we accumulate significantly more games than our opponents accumulate, should we care if our players are efficient? If I play ten guys every night during the playoffs and my opponent plays only five, assuming my players are not completely talentless, I will typically win the countable categories.

By disregarding turnovers and the ratio categories, we can focus our player searches on extremely active players who either touch the ball frequently when they are on the court, or who play a lot of minutes, or both. This thought process leads me to believe that, in H2H leagues, we could do well to disregard turnovers and all ratio categories *for the entire season*. We could focus on winning the countable categories and happily accept the times we win the ratio categories.

Such a strategy could work well in an H2H league where teams simply win or lose every match during the regular season, and cumulative category results *don't* factor into the team's overall standings. Using said strategy, let's assume that for a full season, a manager cumulatively wins an impressive 75% of six non-turnover countable categories (e.g. 3PTM, REB, BLK, ST, BLK, PTS), and the manager wins a reasonable 45% of the remaining ratio or negative categories (e.g. FG%, FT%, TO). This manager's overall win-loss record might be impressive, but his category winning percentage would be only 58.5% (6 x 0.75 + 3 x 0.45).

In the H2H league I joined, cumulative category winning percentages determined our overall position, and a winning percentage of 58.5% would not have been sufficient to earn a first round bye in the playoffs. Because earning a first round bye improves our chances of winning the championship, I'm leaning towards *not* ditching the ratio categories as a full-season H2H strategy if I join this type of league again. In addition, I will likely not ditch turnovers either during the regular season in this type of league until the season unfolds and evidence indicates I can do so without concern. This offseason I'll continue my research in these sensitive areas.

THE OFFSEASON

That's it. That's all I have. Hopefully next season I'll learn more and continue to improve my methods and win my leagues again. Then again, maybe this year was a fluke. Maybe I'll flop in the upcoming season and finish in last place in all my leagues! Regardless of how I perform next year, I'm sure I'll have *fun*! ☺

When I started this project, I didn't know who to write for. I wanted to offer something for *all* fantasy sports managers. The playbook is complete now, and I'll share my thoughts:

If only one person who has never joined a fantasy sports league is encouraged to do so as a result of reading this manual…

If only one fantasy sports manager who reads this book implements some of my suggestions and improves his performance as a result…

If only one person was entertained from reading this playbook…

…then I will consider this project a success.

Thanks again to everyone who helped make this book possible, and thank *you* for reading!

THE PLAYBOOK

PRESEASON

- Select your leagues. My choice is Yahoo! For the 2009-10 season, I'll join at least one roto league and at least one H2H league.

- Join leagues that do not limit your transactions.

- Join small leagues (up to 100 draft picks) or standard-sized leagues (up to 150 draft picks). Avoid very deep leagues (200+ draft picks).

- Join two or three leagues if you have the time. Constantly analyzing one league might lead to burnout or boredom. Joining too many leagues can spread you too thin.

- Allow yourself at least three hours per week to adequately monitor each league and to research player performances within that league. If you join a league with season-ending playoffs, allow thirty minutes *every day* during the playoffs.

THE DRAFT

- Don't sweat the draft.

- Follow the experts' draft advice. My choice is Rotoworld's Draft Guide.

- The day before the draft, print your customized cheat sheet. At your league's website, pre-rank every player who is expected to hold value in that league. For example, if you join a league with 12 managers and each manager will draft 13 players, pre-rank at least 156 players.

- Don't auto-draft. Participate in live drafts only!

- Immediately following the draft, review your team and look for obvious weaknesses. Don't hesitate to enter your request for suitable replacements from the waiver wire.

- During the days after the draft, check your league for activity. Identify strong owners in your league and monitor their actions.

OPENING DAY

- Go to Rotoworld's Season Pass website. In the 'Roto Direct' section of the site, enter your players from all your leagues and enter your email address so Rotoworld will mail you feature articles about the NBA and timely information about your players.

- Peruse the entire Season Pass site. There are many excellent features and articles that help guide managers throughout the NBA season.

THE SPREADSHEET

- Using all your league's categories, determine the minimum cumulative zTotals that should define the three performance tiers for the league. I offer a procedure in the appendix of this manual to help with this task.

- Every week during the season, record the Season zTotals, 45-day zTotals, and 10-day zTotals for all the players on your rosters using all your league's categories. Use a separate worksheet for each team.

- Use the Expected zTotal formula to help you evaluate your players and to predict their future performance levels. Adjust the formula as you deem appropriate.

- Maintain players on your rosters whose Expected zTotals meet your minimum requirements for Tier 1 (gold) or Tier 2 (green) performance levels.

FREE AGENT SEARCH

- ◆At least once each week, use the Ziguana Player Ranks report to search for quality free agents in your league. Be sure to click 'Update Rosters' on your 'My Leagues' page before you search.

- Select the categories you wish to focus on.

- ◆Filter for free agents only. [This feature is currently available for Yahoo! and ESPN leagues only.]

- If the season is at least six weeks in, restrict your search to the last 45 days.

- Open a new web browsing tab and run the same free agent report. This time use stats from only the last ten days.

- When analyzing these two reports, focus on players with very high zTotals during both time periods.

- When you discover a free agent who has been playing very well for the past six weeks, and he has maintained or exceeded his high performance level during the past ten days, research the player further and consider adding this player to your roster.

- Before you add/drop players, compare the upcoming schedules of both players. If the free agent's team is scheduled to play significantly fewer games in the next several days than your current player's team is scheduled to play during the same time period, consider waiting before you make the transaction.

THE PLAYOFFS

- Accumulate as many games played as possible.

- In daily leagues, determine which free agents are playing well now *and* whose teams are scheduled to play the next day.

- At the Yahoo! league site, go to the 'Player List' search page.

- At the stats pull-down field, select 'Ranks.'

- Sort the players three times. First sort them by their 'Season' ranks, then sort them by their 'Last Month' ranks, and finally sort them by their 'Last Week' ranks.

- After each sort, identify every player listed on the first screen (top 25) who is scheduled to play the following day. Add each of these players to your Yahoo! watchlist by clicking on the blue flag beside his name.

- ♦At Ziguana, update your rosters.

- At the 'Player Ranks' page, select all your league categories except turnovers and personal fouls. Also consider disregarding the ratio categories. Run the report.

- Filter for 'My Watchlist' players.

- Restrict your search to the last ten days.

- Open a second browsing tab, run the same report and limit the time frame to only the last few days.

- Determine which players on your roster are not scheduled to play the following day. Consider replacing these players with suitable free agents who *are* scheduled to play the following day.

- Do not drop your very best players unless you encounter a desperate situation.

- Before you add/drop, check the news for your target free agent and confirm he is expected to play with the team the next day. Avoid recently injured players and players who might be suspended.

- Repeat the procedure every day during the playoffs.

Appendix

									SELECTED CATEGORIES			
								TEAM NAME	1.31	1.92	2.52	2.04
ACQ. DATE	DAYS OWNED	PG	SG	SF	PF	C	R W	PLAYER	SEASON zTotal	45 DAY zTotal	10 DAY zTotal	EXPECTED zTotal
29-Oct	123			1			•	PLAYER 01 *	9.03	9.44	9.86	9.53
29-Oct	123				1	1	•	PLAYER 02 *	4.72	3.36	3.42	3.51
1-Feb	28		1	1			•	PLAYER 03	(2.21)	1.90	5.87	2.68
29-Oct	123	1		1			•	PLAYER 04	2.10	2.71	2.59	2.61
29-Oct	123			1			•	PLAYER 06	3.62	2.42	2.08	2.44
14-Dec	77				1	1	•	PLAYER 05	0.52	2.68	2.30	2.35
20-Feb	9				1	1	•	PLAYER 07	(1.20)	1.15	1.95	1.16
14-Nov	107	1		1			•	PLAYER 08	(0.72)	0.46	2.51	0.96
15-Jan	45		1				•	PLAYER 09	1.01	1.06	0.41	0.86
14-Jan	46			1			•	PLAYER 10	0.80	0.23	1.61	0.70
1-Mar	0				1		•	PLAYER 11	0.06	0.52	0.56	0.49
1-Dec	90		1				•	PLAYER 12	(0.79)	0.29	1.26	0.47
14-Feb	15					1	•	PLAYER 13	0.10	(1.30)	(1.61)	(1.25)

CAN'T CUT *

TRADE OFFERED

TRADE PENDING

ATTENTION REQUIRED

PG	2
SG	7
SF	4
PF	4
C	4

TIER 1 MINIMUM zTOTAL	2.50
TIER 2 MINIMUM zTOTAL	0.50
TIER 3 MINIMUM zTOTAL	(1.00)

Cells A1:I2 This merged cell displays your team name. If you join two or more leagues, you should create a separate worksheet for each of your teams within the workbook. You might also change the name of each worksheet *tab* to match your team's name. These visual references are useful when you monitor multiple teams and leagues.

A4:A16 Acquisition date. Enter the date you acquire the player.

B4:16 Days owned. These cells are formula driven and the values update automatically every time you open the file. Excel calculates the difference between the player's acquisition date and the current date. When you know how long you've owned a player, you can determine the player's performance level for the exact time frame you've owned him.

C4:G16 Player position. Enter a '1' in the appropriate column for every position the player is eligible to play in your league.

H4:H16 'RW' is short for Rotoworld. After you acquire a new player, enter a symbol in this column to remind yourself that you have added this new player to your customized player list at Rotoworld's Season Pass website. For players you release, you should remove them from your Rotoworld player list, but first confirm the player is not still a member of one of your *other* teams.

I4:I16 Enter each player's name. For sorting purposes, I prefer to enter the player's last name first, followed by a comma and the player's first name.

J4:L16 This section is the heart of the spreadsheet. Once every week during the season, I recorded each player's Season zTotal, his 45-day zTotal and his 10-day zTotal for all the categories my league scored. The quickest way to obtain this information is by using Ziguana's 'My Teams' page. The process requires less than five minutes per team.

These cells are conditionally formatted to change color based on the value you enter in these cells, and based on the value you enter in cells M19:M21. Gold cells visually represent the highest performance level, or the "gold standard" of performances. I refer to this highest performance level as the 'Tier 1' level.

Green cells represent good, or respectable, performance levels, and I refer to this level as the 'Tier 2' performance level. Grey cells represent mediocre, or 'Tier 3,' performance levels, and Red cells indicate the player's performance level for the selected time period is unacceptable.

By viewing these colorfully illustrated data from left to right for each player, it's easy to see the performance trends and to determine who is performing well and who is underperforming.

M4:M16 Expected zTotal. These cells are formula driven and they are also conditionally formatted to change color to gold, green, grey and red. The Expected zTotal is my way of valuing any player in the NBA, and it's my attempt to predict a player's future performance level. The current formula calculates an expected value based on the following criteria:

Season zTotal	= 10% of the Expected zTotal
45-day zTotal	= 60% of the Expected zTotal
10-day zTotal	= 30% of the Expected zTotal

I adjusted this formula numerous times during the season in my attempt to predict future performance levels, and I encourage you to experiment with different formulas as well. With that said, if you use the formula shown above, you'll be fine. It's more important to stay active and regularly review your player's performances than it is to spend too much time trying to determine the "perfect" performance predicting formula.

ROW 3 Row 3 is set to auto filter. You can experiment with this auto filter for each column of data, but you will certainly use the 'Sort Ascending' and 'Sort Descending' features of this useful spreadsheet tool. By default, I sort my players in descending order by Expected zTotal (column M).

J1:M1 These cells are formula driven and they display the average of all your players' zTotals shown in columns J through M, respectively.

These average zTotals provide a useful indicator of your well your team, as a whole, has performed for the given time periods. These cells are also conditionally formatted to change color to gold, green, grey or red. One of your goals should be to maintain players with zTotals high enough to turn these cells gold or green.

J2:M2 Use this merged cell to display the names of the categories you are analyzing. I narrowed my category searches as the season progressed, but for most of the season, I analyzed players using all the categories in the league. I typically entered "All Categories" in this cell. Keep in mind the fewer the categories you analyze, the lower the cumulative zTotals will be.

E19:F23   Cells E19:E23 are formula driven and they calculate the sum of the values in cells C4:G16, respectively. The values displayed in cells E19:E23 indicate the total number of players on your roster who are eligible to play each listed position in your league, as shown in the adjacent cells F19:F23. Cells E19:E23 are conditionally formatted to turn pink if the calculated value drops below three.

It's helpful to maintain a balance of positions during the season. Few things were more frustrating than the times I was forced to bench quality ball players because I lacked a sufficient number of players who were eligible for a given position. I learned that if I maintained players who, collectively, covered each of the five positions at least three times, I could usually avoid this benching headache.

It doesn't hurt to have an *abundance* of players who are eligible to play any given position (see the seven shooting guards in the screenshot above), but it certainly hurts to have too few. By maintaining a minimum of at least three players who were eligible to play each position, I was able to accumulate the maximum number of games allowed in all my leagues for every position.

I19:I22 These four cells comprise a legend used for reference.

I19 Can't Cut player. It helps to identify players whom you cannot drop from your roster. In cells I4:I16 where I enter the players' names,

I identify each 'Can't Cut' player by entering an asterisk (*) after his name.

I20 Trade Offered. If I offered a trade for any of my players, I preferred to remind myself of it when viewing this file. In cells I4:I16, I underline the name of any player for whom I had open trade offers.

I21 Trade Pending. If my trade offer was accepted, or if I accepted a trade offer from another manager, in cells I4:I16 I *italicized* the name of the soon-to-be-traded player.

I22 Attention Required. It's helpful to visually identify players who require special attention. In cells I4:I16 I manually colored the cell pink for any player who needed to be closely monitored. There can be many reasons to carefully monitor a player, the most common of which being personal injuries.

M19:M21 These cells are critical and they require manual numerical inputs. Cell M19 is colored gold and this cell displays the minimum cumulative zTotal required to be considered a Tier 1 performance in your league using all your league's categories. M20 This cell is colored green and it displays the minimum cumulative zTotal required for a Tier 2 performance using all the league's categories. Cell M21 is colored grey and it displays the minimum zTotal required for Tier 3 performances.

To determine the minimum cumulative zTotals that define the three performance tiers in your leagues, first determine the total number of NBA players who can *start* each night in your league. This figure should be calculated as such: Number of teams in your league x Number of allowable daily starting players per team (not the total number of players on each team). Divide this figure by three to determine the number of players who qualify for each of the three performance tiers.

Here's an example of this tier-defining process using commonly used figures in Yahoo! leagues: 12 teams in league x 10 daily starting players per team = 120 daily starting players. Then, 120 / 3 = 40 Tier 1 players, 40 Tier 2 players, and 40 Tier 3 players.

To begin the season, go to the Ziguana Player Ranks page, select all the categories from your league and rank all players using *last season's data*. [After the current season is at least six weeks in, you should adjust these tier-defining zTotals to reflect the current season's data.]

Scroll down to locate the three players whose *zRanks* would place them at the very bottom of Tier 1, the bottom of Tier 2, and the bottom of Tier 3. Record the cumulative zTotal for each of these three players. In the example above, you would record the cumulative zTotal for the player whose zRank is 40, for the player whose zRank is 80, and for the player whose zRank is 120.

As a reminder, I only assign value to the number of players who can *start* any given day in my leagues. You can value fewer players if you prefer, but doing so will increase the minimum required zTotals for each performance tier. Consequently, you might be frustrated if you cannot retain players who consistently achieve impractical performance standards. Conversely, if you value *more* players, your standards might be too low.

To continue my example, in one of my nine-category leagues, using data from the 2008-09 NBA season, the player with a zRank of 40 had a full-season zTotal of 2.52. For convenience, I rounded this figure to 2.50 and I entered this number in cell M19 of the spreadsheet. The player with a zRank of 80 scored a season zTotal of 0.48. I rounded this figure to 0.50 and I entered the value in cell M20. The player with a zRank of 120 scored -1.02 and I entered -1.00 in cell M21.

By comparison, in my *eight*-team, nine-category league, I valued only 80 players for any given search (8 teams * 10 starters per team), and about 27 players (80 / 3) qualified for each of the three performance tiers. Using this past season's data, the player with a zRank of 27 had a season zTotal of 3.48 using the categories from my league. If I join an 8-team league this fall that scores the same categories, I will begin the season using 3.50 as my Tier 1 minimum performance level.

The player with a zRank of 54 realized a season zTotal of 1.52. Again, if I join a similar league this October (same number of teams, same categories), I will enter 1.50 in cell M20 as my Tier 2 minimum performance standard. The player with a zRank of 80 scored 0.27 this past year, and I will use 0.25 in cell M21 as my Tier 3 minimum performance standard if I join a similar league.

You'll discover z-scores and zTotals change daily, and they can change significantly over short periods of time. However, after following the procedure outlined above, you won't need to constantly adjust these performance standards (cells M19:M21). I do recommend, however, refreshing these values a few times during the season. After about six weeks into the season, the zTotal tier requirements won't change significantly, but it's a useful exercise to occasionally update these figures. Doing so helps us gain a better understanding and appreciation for zTotals, and it ensures we are analyzing performance levels using current data.

The next screenshot shows another portion of the spreadsheet. This section displays player- and team-related information that is used to determine the number of valuable fantasy players in your league. This section also calculates useful data concerning the time remaining during the fantasy season.

	I	J	K	L
40				
41		**TODAY'S DATE**	**1-Mar**	
42				
43		**# OF TEAMS IN LEAGUE**	**12**	⇐
44		**# OF STARTERS / TEAM**	**10**	⇐
45				
46		**# TIER 1 PLAYERS IN LEAGUE**	**40**	
47		**# TIER 1,2 PLAYERS IN LEAGUE**	**80**	
48		**# TIER 1,2,3 PLAYERS IN LEAGUE**	**120**	
49				
50		**FANTASY SEASON END DATE**	**15-Apr**	⇐
51		**DAYS REMAINING**	**45**	
52		**WEEKS REMAINING**	**6.4**	
53		**MONTHS REMAINING**	**1.5**	
54		**~ GAMES REMAINING**	**21.7**	
55		**% SEASON REMAINING**	**26%**	

Cell K41 Today's date. This cell is formula driven. It automatically refreshes the current date each time you open the file. This value is used to compute the number of days you own each player on your roster, which I described earlier.

K43 Number of teams in league. Manually enter the total number of competing managers in your league.

K44 Number of starters per team. Manually enter the total number of ball players that are allowed to start on each team on any given night.

K46:K48 Number of tiered players in your league. These cells are formula driven. K46 displays 1/3 of the total number of starters in your league. This figure represents the number of Tier 1 players in the league for any given player search.

K47 Displays 2/3 of the total number of starters in your league. This figure represents the cumulative total of Tier 1 and Tier 2 players in the league.

K48 Displays the total number of starters in your league. This figure represents the cumulative total of all tiered players in the league. This figure also represents the total number of ball players in your league who hold considerable value for any given player search.

K50 Fantasy season end date. Manually enter the last day's date of fantasy season play. For H2H leagues, you might consider entering the ending date of the regular season.

K51:K55 These cells are formula driven, and they update automatically every time you open the file.

K51 Days remaining. This cell displays the difference between 'Today's Date' and the 'Fantasy season end date.'

K52 Weeks remaining. Displays the number of days remaining divided by seven.

K53 Months remaining. Displays the number of days remaining multiplied by 365 and divided by 12.

K54 ~ Games remaining. Displays the number weeks remaining multiplied by 82 (the number of games each team plays in a season) and divided by 24.3 (the number weeks in a season).

K55 Percent season remaining. Displays the number of games remaining divided by 82, and the display is expressed as a percentage.

The next two screenshots display optional, but highly suggested, portions of the spreadsheet. I updated these sections every week. The first of these screenshots is copied from one of my rotisserie leagues. It displays a summary of my overall league rank, as well as my league standings in each category. The final screenshot is copied from my H2H league. It also displays a summary of my overall rank, and it includes my results for each category from each week's match.

	27-Dec	3-Jan	10-Jan	17-Jan	24-Jan	31-Jan	7-Feb	14-Feb	21-Feb	28-Feb	7-Mar	14-Mar	21-Mar	28-Mar	4-Apr	11-Apr
FG%		7.0	6.0	6.0	6.5	6.0	7.0	7.0	7.0	7.0	7.0	7.0	7.0	7.0	7.0	7.0
FT%		3.5	4.0	4.0	4.0	4.0	4.0	4.0	4.0	4.0	4.0	4.0	4.0	3.5	4.0	3.0
3PTM		6.0	6.0	6.0	6.0	5.0	6.0	6.0	6.0	6.0	6.0	6.0	6.0	6.0	6.0	6.0
PTS		6.5	7.0	7.0	7.0	7.0	7.0	7.0	7.0	7.0	7.0	7.0	8.0	8.0	8.0	8.0
REB		7.0	7.0	7.0	7.0	7.0	8.0	8.0	8.0	8.0	8.0	8.0	8.0	8.0	8.0	8.0
AST		7.0	8.0	7.0	7.0	8.0	8.0	8.0	8.0	8.0	8.0	8.0	7.0	7.0	7.0	7.0
ST		7.0	7.0	7.0	7.0	6.0	6.0	7.0	7.0	7.0	7.0	7.0	7.0	8.0	7.0	8.0
BLK		6.0	6.0	6.0	6.0	6.0	5.0	6.0	6.0	7.0	6.0	6.0	8.0	8.0	8.0	8.0
A/T		5.0	6.0	6.0	6.0	6.0	6.0	6.0	6.0	6.0	6.0	6.0	6.0	6.0	6.0	6.0
Δ		2.0	2.0	(1.0)	0.5	(1.5)	1.0	4.0	(1.0)	1.0	(1.0)	1.0	2.0	(1.5)	0.5	(1.0)
TTL		55.0	57.0	56.0	56.5	55.0	56.0	60.0	59.0	60.0	59.0	60.0	62.0	60.5	61.0	60.0
EFF		76%	79%	78%	78%	76%	78%	83%	82%	83%	82%	83%	86%	84%	85%	83%
RANK		3	2	2	2	2	2	1	1	1	1	1	1	1	1	1

	# WKS COMPL	WEEK NO. →														
	21	21	20	19	18	17	16	15	14	13	12	11	10	9	8	7
FG%	0.389	1	0	0	0	1	1	0	0	1	0	0	1	0	0	1
FT%	0.444	0	1	1	0	0	0	1	0	0	0	0	1	0	1	1
3PTM	0.944	1	1	1	0	1	1	1	1	1	1	1	1	1	1	1
PTS	0.833	1	1	1	1	1	1	1	1	1	1	1	1	1	1	0
REB	0.750	1	1	1	0	0.5	1	1	1	0	1	1	1	1	1	1
AST	0.861	1	1	1	1	0.5	1	1	1	1	1	1	1	1	1	1
ST	0.889	1	1	0	1	1	1	0	1	1	1	1	1	1	1	1
BLK	0.833	1	1	1	1	1	1	1	1	0	1	1	1	1	1	1
TO	0.222	0	0	1	1	0	0	0	0	1	0	0	1	0	0	0
EFF.	0.685	7.0	7.0	7.0	4.0	5.0	7.0	6.0	6.0	6.0	6.0	6.0	9.0	6.0	7.0	7.0
		0.78	0.78	0.78	0.44	0.67	0.78	0.67	0.67	0.67	0.67	0.67	1.00	0.67	0.78	0.78
RANK	1	1	1	1	1	1	1	1	1	1	1	1	1	1-2	3	2

86

Cell values that display positive results (i.e. a category I improved in or a category I won) are conditionally formatted to turn green. Cell values that display negative results (i.e. a category I worsened in or a category I lost) are conditionally formatted to turn red. I can't honestly say that analyzing this weekly information was critical to my success, but I did, however, use this data to confirm trends.

For instance, at one point in one of my rotisserie leagues, I targeted two categories in which I hoped to improve, and I acquired players who I thought would help me in these areas. By visually tracking my position in each category, I was able to confirm that my strategy was paying off. If I had not regularly recorded this data, I would have been forced to *remember* what my league standings in these categories were before I targeted them for improvement.

In isolation, this example might seem silly because, after all, just how hard is it to remember one or two figures (in this case, beginning category scores)? However, when monitoring *multiple leagues*, it can be challenging to remember and mentally separate historical information from one league to another. Therefore, to prevent overworking my memory cells, I will surely use these portions of the spreadsheet in future seasons.

On these spreadsheets, I also maintained records of all my trade data – trades offered and received, accepted and rejected – which provided useful references when I offered or received additional trades. Next season I'll capture even more information. I have ideas for recording the league-leading stats for each category and for comparing each of these stats to my teams' stats, both as a percentage and as whole figures. Trend analysis in fantasy sports shouldn't be underestimated, and it's my intention to take the analyses – both analytically and visually – as far as my imagination will allow me to. ☺

Reminder: if you would like a copy of this spreadsheet, please visit www.awinningplaybook.com to download the file.

www.ingramcontent.com/pod-product-compliance
Lightning Source LLC
Chambersburg PA
CBHW051208050326
40689CB00008B/1236